written by
Rozanne Lanczak Williams

Contents

What Do You See? 2
Glossary 11
Index 12

Orlando   Boston   Dallas   Chicago   San Diego

www.harcourtschool.com

Look on this leaf.
What do you see?

I see a little egg.
What will it be?

Look on this leaf.
What do you see?

I see a little caterpillar.
What will it be?

Look on this stick.
What do you see?

I see a little pupa.
What will it be?

Look on this flower.
What do you see?

I see a butterfly!
That is what I see.

Fly, fly, butterfly!

Glossary

butterfly

caterpillar

egg

pupa

Index

butterfly, 9, 10
caterpillar, 5
egg, 3
pupa, 7